Economic Migration and Climate Change

The coming invasion

(The connection)

By

Prof. Stephen W. Bradeley BSc (Hons)

COPYRIGHT

Table of Contents

ABOUT ME

Writer profile

My name is Steve Bradeley and I am 66 years old (2022). My name is Steve Bradeley and I am 66 years old (2022). I am an avid motorcycling enthusiast, I embrace life with a diverse blend of passions. As a writer, I try to masterfully weave together narratives that captivate readers and leave them yearning for more.

My adventurous spirit drives me to explore new corners of the world, fueling my love for travel by motorcycle. In addition to these personal interests, I remain actively interested in the realm of politics, continuously broadening my understanding and formulating informed opinions. Overall, I am a multifaceted individual whose experiences and interests make me both an interesting person. I hold a science degree from Staffordshire University.

Introduction

In the 21st century, humanity faces an unprecedented global crisis. Our planet is undergoing drastic transformations due to climate change, causing widespread destruction and upending lives in the most vulnerable regions of the world. As a result, millions of people are left with no choice but to migrate in search of a better future. This book, "Economic Migration and Climate Change: The Coming Invasion", aims to shed light on the intricate connections between climate change and mass migration, particularly from the Southern Hemisphere to the Northern Hemisphere. It will explore how these shifts carry significant social, political, economic, and environmental consequences for both Europe and North America.

The effects of climate change can no longer be ignored. According to the United Nations High Commissioner for Refugees (UNHCR), approximately 21.5 million people are displaced annually due to climate- or weather-related events.

This number is projected to rise exponentially in the coming decades.

Furthermore, experts predict that by 2050, between 150 million and 1 billion people might be displaced as a result of environmental changes - many seeking refuge in countries less affected by climate disasters.

A primary reason for this mass exodus is the increasing frequency of extreme weather events such as droughts, flooding, storms, and heatwaves caused by changes in global temperature patterns. These events disproportionately impact low-income communities in Africa, Southeast Asia, and Central America – causing widespread food shortages, resource conflicts, loss of livelihoods, mass public health crises, environmental degradation, and reduced access to clean water.

As a result of these challenges, climate-induced conflicts are becoming prevalent worldwide. Regions affected by water scarcity are particularly prone to conflicts over diminishing resources – leading to large-scale civil unrest or interstate wars. This book will delve into various case studies that demonstrate the increasing connection between climate change and conflict incitement in vulnerable regions across South Asia and Sub-Saharan Africa.

Europe and North America are and will continue to be profoundly affected by this growing trend of climate-induced migration. The influx of migrants to these regions will put immense pressure on existing infrastructure, social systems, and the environment. This increased demand for resources might lead to both positive and negative economic consequences, impacting job markets, public service expenditure, housing availability, and cultural dynamics.

However, with every challenge comes an opportunity. This book will also explore the potentially beneficial aspects of migration, including its capacity to fill labor gaps in aging societies; promote innovation through enhanced diversity;

foster global collaborations by encouraging cross-border networks and exchange of ideas; and the positive economic impacts of remittances.

Finally, "Economic Migration and Climate Change: The Coming Invasion" will discuss the geopolitical ramifications of this overwhelming migration trend. The issues related to national security, political stability, international cooperation, border control, foreign policy strategies, and their collective impact on the global stage will be thoroughly examined.

As you dive into this book, prepare yourself to embark on an enlightening journey that explores the complex relationship between climate change and human migration. It is our hope that by understanding these connections better, we can inspire meaningful conversations among policymakers and the general public alike – ultimately leading to an informed global response that addresses the root causes of these pressing issues while offering holistic solutions. Together, we can confront the challenges ahead and secure a more sustainable future for generations to come. Now, let's look at this in more detail.

Chapter One

Present situation of economic migration into Europe

———

Economic migration has been a prevalent phenomenon for centuries, with individuals and families across the globe seeking better opportunities in other countries. Europe has been a prime target for these migrants due to the promise of higher living standards, economic development, and social support. However, the landscape of economic migration into Europe has shifted significantly over the past few decades, with an increasing number of individuals arriving not only through legal channels but also by means of dinghies and rafts. This chapter will delve into the present situation of economic migration into Europe, examining pertinent statistics related to travelers' method of arrival.

Arrival by Sea: A Risky Endeavor

One prevalent mode of entry for migrants seeking new horizons is via sea, specifically through the use of small boats such as dinghies and rafts.

This method poses immense risks to passengers due to overcrowding and treacherous routes. Nonetheless, thousands embark upon this journey each year.

According to data from the United Nations High Commissioner for Refugees (UNHCR), in 2020 approximately 95,000 refugees and migrants arrived in Europe through Mediterranean crossings. Greece was the primary destination, receiving around 40% of these arrivals. Italy came in second with slightly more than 34,000 individuals. Spain followed with around 17% of total arrivals.

Crossing the English Channel

The English Channel serves as another crucial route for migrants attempting to reach mainland Europe and UK territory alike. The number of people attempting this crossing has been on the rise over recent years due to increased restrictions on land-based routes.

IN 2020 ALONE, MORE than 9,000 people crossed the English Channel by small boat—almost four times the number recorded in 2019 when approximately 2,300 people made that perilous journey. The majority arrived from France, packed into unsafe dinghies or rafts, demonstrating the desperate measures many economic migrants are willing to take to secure a better life.

Factors Influencing Economic Migration

It is crucial to delve into the reasons behind this influx of economic migration into Europe. Existing literature has attributed this trend to various interconnected factors, including:

1. Economic disparities: Significant income gaps between migrants' countries of origin and European nations compel individuals to seek work and a higher quality of life outside their homeland.

2. Political instability: Conflict, persecution, and human rights abuses in places like Syria, Afghanistan, and Eritrea have resulted in numerous refugees seeking safety in European territories.

3. Demographic transition: Aging populations in certain European countries offer new job opportunities for migrants seeking employment.

Challenges and Future Implications

Economic migration into Europe presents both challenges and opportunities for receiving countries. On one hand, it has led to cultural enrichment, increased diversity, and the filling of labor shortages. On the other hand, rising numbers of economic migrants have strained integration resources.

European nations must address the issue from multiple angles: enhancing cooperation among nations, managing borders more effectively without jeopardizing human rights, investing resources into integrating newcomers, and facilitating legal pathways for migration. In turn, these measures may help reduce reliance on dangerous sea-crossing methods such as dinghies and rafts while providing a fair and humane approach for those genuinely seeking a better life.

8

Economic migration into Europe is a complex reality that has seen thousands of people risking their lives by crossing seas in small boats. Discerning the factors driving these individuals and addressing them is critical not only to alleviate suffering but also to foster stability within host nations. By understanding these statistics and developing policies accordingly, Europe can seek sustainable solutions to deal with current economic migration trends with humanity at its forefront.

Chapter Two
Present European policies

In Europe, economic migration has become a significant issue in recent years. As a result, various policies have been implemented to address these concerns. This chapter will examine the present policies of the European Parliament and the UK Government on economic migration into Europe. We will also discuss key political figures and their attitudes towards these policies.

I. THE EUROPEAN PARLIAMENT Policies on Economic Migration

A. The EU Blue Card Directive

In response to the increasing demand for skilled labor, the European Union introduced the Blue Card Directive in 2009 as a work permit for highly skilled non-EU nationals. This initiative is designed to attract qualified workers from outside the EU by offering incentives such as streamlined visa processes, social benefits, and family reunification rights.

B. The Asylum, Migration, and Integration Fund (AMIF)

The AMIF aims to support member states in managing migration effectively. Established in 2014, it provides €3.137

billion in funding for initiatives related to asylum seekers, integrating legal migrants, and addressing irregular migration.

C. Key Figures and Their Attitudes

1. Ursula von der Leyen: As President of the European Commission, von der Leyen has advocated for a balanced approach to migration that emphasizes both responsibility and solidarity among EU member states.

2. Ylva Johansson: The current European Commissioner for Home Affairs, Johansson is responsible for formulating policies relating to migration and asylum within the EU.

II. The Present UK Government Policies on Economic Migration into Europe

A. The Post-Brexit Points-Based Immigration System

Following Brexit, the UK has introduced a new points-based immigration system that prioritizes skills over nationality. This system evaluates applicants based on factors like qualifications, English language proficiency, and salary to determine eligibility for work visas.

B. THE IMMIGRATION Skills Charge (ISC)

To enourage employers to invest in domestic workforce, the UK Government has implemented the ISC. This fee is levied on businesses sponsoring skilled non-EEA workers, directing the collected funds towards funding apprenticeships and improving workforce capabilities.

C. Key Figures and Their Attitudes

1. Boris Johnson: As Prime Minister of the UK, Johnson focused on creating an immigration system that favors talent and skills, regardless of applicants' countries of origin.

2. Priti Patel: As Home Secretary, Patel implemented policies designed to control UK borders and prioritize the nation's needs when welcoming migrants.

Both the European Parliament and the UK Government have implemented policies aimed at addressing economic migration into Europe. With a focus on skills and international cooperation, their goal is to create a system that benefits European economies while providing opportunities for those seeking to migrate for better prospects. Unique political figures also play essential roles in shaping these policies, influencing their effectiveness and overall impact on migration trends.

Chapter Three
The Dark Side

———

Economic Migration and the English Channel Crisis: The Impact of Public Opinion and Far-Right Involvement

Individuals and families venture from their home countries in search of better opportunities, often brought about by economic necessity. The English Channel, as a significant pathway between continental Europe and the United Kingdom, has become a focal point for economic migration. This chapter delves deep into this phenomenon, exploring how it manifests in daily boat arrivals to England's shores and how it has impacted public opinion. We will discuss the evolving attitudes towards those seeking better opportunities, the rise of far-right involvement, and the implications of racism on this crisis.

The English Channel as a Route for Economic Migration

The strategic geographic position of the English Channel makes it a significant migratory route for those seeking economic refuge in the United Kingdom. The increase in migrants crossing by boat has triggered heated debates about Britain's ability to accommodate such influxes and whether these migrants should be permitted to enter or reside within its borders.

Changing Public Opinions on Boat Arrivals

In recent years, complex global issues have contributed to increased economic migration via the English Channel. News coverage on daily boat arrivals has fueled public discourse on immigration policy while also shaping perceptions about migrants. While certain segments of British society may express empathy towards those seeking refuge from hardship overseas, others see these migrants as burdensome or threatening.

THE RISE OF FAR-RIGHT *Sentiment and Racism*

Amidst changing public attitudes, far-right political parties have capitalized on concerns over economic migration as a way to fuel anti-immigrant sentiment and consolidate power. Many of these parties push for restrictive immigration policies, asserting that migrants take away resources from native-born citizens and present risks to national security.

Far-right activists have been known to gather around ports along the English Channel, where they harass migrants as they attempt to enter the country. Targeting both asylum seekers and those seeking economic opportunities, these confrontations reflect the growing reach of xenophobia in tandem with negative attitudes towards economic migrants.

The Impact of Racism on the Economic Migration Crisis

Racism plays a critical role in shaping public reactions to the English Channel crisis. Prejudice and discrimination against those arriving on boats often arise from presumptions that these individuals hold cultural values incompatible with British society. Racial profiling and stigmatization marginalize these newcomers, potentially exacerbating existing socioeconomic inequalities and making successful assimilation more challenging.

The English Channel migration crisis raises significant questions about how societies should respond to economic migration in an age of increasing global interconnectedness. In England, public opinion has proven pivotal in determining the fate of those arriving across its borders. As attitudes evolve and far-right sentiment gains more traction, future policies may determine not only who is allowed into England but also how migrants will be approached or integrated within their new homes. The implications of this ongoing debate extend far beyond the shores of the English Channel, confronting policymakers worldwide with difficult decisions on balancing empathy for migrants with concerns over their potential impact on host nations.

Chapter Four
The Perilous Journey Across the Mediterranean Sea and English Channel

Economic migration has emerged as one of the most pressing global challenges as people from impoverished or war-torn nations leave in search of stability, security, and prosperity. This chapter delves into a critical aspect of economic migration—the dangers faced by migrant families as they embark on perilous journeys across the Mediterranean Sea and the English Channel in small, overcrowded boats.

The Dangerous Crossings

Every year, thousands of migrants attempt the treacherous journey across the Mediterranean Sea and the English Channel. These waters have claimed many lives as desperate people brave powerful currents in small, makeshift boats often overloaded with passengers without life vests or other safety equipment.

THE ABSENCE OF LEGAL safe passages for these migrants forces them to resort to irregular crossings via smugglers who prioritize profits over safety. The lack of proper protection from the elements, inadequate food and water supplies, and minimal navigational tools often lead to tragic consequences. Migrants suffer from dehydration, hypothermia, and motion sickness or even lose their lives when the boats capsize or get caught in storms.

In addition to the physical dangers they face during these crossings, migrants often face relentless abuse at the hands of smugglers. In an industry fueled by desperation and vulnerability, exploitation is rampant. Migrants are sometimes held hostage and subjected to torture if their families do not pay extra fees demanded mid-journey.

Criminal Gangs Behind These Trips

The business of smuggling migrants is a lucrative one for criminal networks who exploit these vulnerable individuals for profit. Organized crime syndicates operate highly sophisticated structures that facilitate illegal border crossings. They employ various means to move migrants through Africa to Europe or through Central and South American countries to North America.

These criminal networks often maintain contact with corrupt officials who turn a blind eye to their activities in exchange for bribes. Bribing officials also enables the gangs to be one step ahead of law enforcement, making it difficult for the latter to dismantle these illicit operations.

Criminal gangs capitalize on the desperation of migrants by offering a range of services at exorbitant prices. These services might include false documentation, which can cost thousands of dollars, and passage across borders through water or over land. Often, these costs are borne by families who are forced into debt in order to finance their loved ones' journeys.

Economic migration has become a critical issue as individuals continue to take extreme risks in search of better opportunities. The grave dangers faced by them during sea voyages, especially in the Mediterranean Sea and the English Channel, highlight the urgency for policymakers to implement safe and legal pathways for migration. Moreover, concerted efforts must be made to dismantle criminal networks that profit from these perilous journeys and ensure that justice reaches those who have been cruelly exploited along the way.

Chapter Five

Economic Migration and the United Kingdom in 2023

Present UK Policies on Economic Migration

The UK has undergone substantial policy reform since the Brexit vote in 2016. By 2023, the country's departure from the European Union (EU) has significantly impacted its approach to immigration policy. The implementation of a points-based system seeks to attract highly skilled professionals worldwide while maintaining control over the flow of unskilled labor.

Under the new Points-Based Immigration System (PBIS), migrants are evaluated based on their professional experience, skills, education, salary level, and ability to speak English. These criteria are designed to ensure that only those who can contribute to the UK economy and social fabric are awarded visas. In addition to this policy, 2023 sees an increased focus on attracting innovative entrepreneurs and start-ups to boost economic growth and job creation.

Changing Public Attitudes Towards Migrants

The past decade has seen fluctuations in public sentiment toward migration in the UK, with attitudes being shaped by factors such as political rhetoric, economic concerns, and perceptions of cultural identity. However, over time there has been a growing appreciation for the role that migrants play within national economies.

In 2023, public opinion appears to be more supportive of skilled migrants contributing positively to society rather than perceiving them as a potential source of competition for resources or employment opportunities. This shift in attitudes can be attributed to campaigns that promote cultural diversity, awareness of the economic benefits of migration, and efforts to counteract misconceptions surrounding the topic.

UK Government Policies to Deal with Migration

Recognizing the complex challenges posed by economic migration, the UK government has introduced a variety of policies and initiatives aimed at managing its implications in 2023. These include:

1. Strengthening Borders: To maintain control over migration and ensure security, the UK government has invested in advanced border technologies, increased personnel, and enhanced international cooperation.

2. Integration Strategies: Addressing the integration of migrants into British society is crucial to fostering social cohesion and reducing tensions between communities. Initiatives focusing on language support, access to essential services, and community engagement have been introduced.

3. Workforce Development: To maximize the potential of both natives and migrants within the labor market, investment in

education and training programs has been prioritized in recent years.

4. Regional Migration Partnerships: Acknowledging that economic migration affects different regions uniquely, localized policy development is encouraged to address specific challenges and opportunities.

By embracing forward-thinking policies that balance control over borders, integration initiatives, and investment in human capital development, the UK aims to create a sustainable framework through which economic migration can be managed effectively in 2023 and beyond.

The evolution of UK policies on economic migration reflects changing public attitudes and an increasing focus on harnessing the potential of talented individuals already residing within its borders or those looking to settle there in search of opportunities. Globalization continues to influence the landscape of economic migration; adapting accordingly will be key for the UK's future success and stability.

Chapter Six

Impact and Consequences of Climate Change on the Southern Hemisphere

---

Impact and Consequences of Climate Change on the Southern Hemisphere

Climate change poses significant threats to our planet's ecosystems, populations, and economies. The Southern Hemisphere, which includes South America, Africa, Australia, Antarctica, and the southern parts of Asia, is experiencing significant consequences of this global crisis. This chapter will delve into the impacts and consequences of climate change on these regions, illuminating country-specific information and detailing recent natural disasters related to this existential threat.

SECTION I: SOUTH AMERICA

1. Brazil - Climate-related phenomena like rising temperature and drastic precipitation changes have led to an increase in fires in the Amazon rainforest. In 2019 alone, there were over 80,000 recorded fires.

2. Argentina - Floods in northeastern Argentina in 2022 resulted in loss of lives and immense damage to infrastructure.

3. Chile - The melting of Andean glaciers poses risks to freshwater reserves for millions of people due to reduced water storage capacity.

Section II: Africa

1. Madagascar - As one of the world's most vulnerable countries to climate change, Madagascar has been hit by cyclones intensified by changing climatic patterns with dramatic effects on agriculture.

2. South Africa - Droughts are becoming more severe and frequent due to reduced rainfall patterns; cities like Cape Town have endangered water supply due to low reservoir levels.

3. Mozambique - Flooding from Cyclone Eloise (2021) devastated homes, displaced thousands, and destroyed vital infrastructure disrupting daily life for millions.

Section III: Australia

1. Bushfires ravaging Australian landscape - The catastrophic bushfire season of 2019-2020, known as the "Black Summer," witnessed a destruction of habitats, loss of property, and a significant impact on wildlife populations.

2. Great Barrier Reef coral bleaching - High sea temperatures have caused extensive coral bleaching in the Great Barrier Reef since 2016.

SECTION IV: ANTARCTICA

1. Ice sheet and glacier melting - As a result of warming temperatures, Antarctica's ice sheets are melting at an accelerated rate, leading to a significant global sea level rise.

Section V: Southern Asia

1. Pakistan - The 2010 floods in Pakistan, which affected nearly 20 million people, were exacerbated by climate change-driven extreme rainfall.

2. India - Increasing heatwaves in many regions of India have caused significant fatalities among vulnerable populations.

The Southern Hemisphere is experiencing the adverse effects of climate change more than ever before. As this chapter has discussed, these consequences range from exacerbated wildfires to destructive floods and threaten the livelihoods of millions of people across the globe.

It is crucial for nations to work collaboratively towards mitigation and adaptation measures that protect ecosystems, communities, and future generations from the devastating impacts of climate change

Chapter Seven

Impact and Consequences of Climate Change and Rising Temperatures

———

Climate change is an unequivocal phenomenon occurring at a global level with far-reaching implications for the environment, human populations, and ecosystems. This chapter will delve into the various impacts and consequences of climate change and rising temperatures on diverse sectors, including agriculture, water resources, health, biodiversity, economy, and social systems.

1. Agriculture and Food Security

Increased temperatures and changes in precipitation patterns will have serious consequences for crop growth, yield, and land suitability. The predominant effects include:

- Reduced agricultural productivity due to heat stress in both crops and livestock

- Decreased water availability for irrigation

- Shifts in the geographical range of crops

- Increased incidence of pests and diseases

- Greater risk of crop failure due to extreme weather events

2. Water Resources

The availability of freshwater resources is closely linked to climate change due to changes in precipitation patterns and glacier melt. Some implications are:

- Reduced water storage in reservoirs, lakes, and snowpacks

- Lower groundwater recharge rates from decreased snowmelt runoff

  – Increased evaporation rates from rising temperatures

- Higher likelihood of droughts and water scarcity

3. Health

The impacts of climate change on human health are both direct, such as through exposure to extreme weather events or thermal stress, and indirect through disruptions in ecosystems that support life. Major health-related consequences include:

- Increased morbidity and mortality from heatwaves

- Greater vector-borne diseases transmission

- Reduced air quality leading to respiratory illnesses

- Declining nutritional quality of food sources

- Proliferation of waterborne illnesses due to flooding

4. Biodiversity Loss

Ecosystems are under strain from climate-related threats that result in a loss of biodiversity:

- Altered lifecycle events, such as breeding or migration timings

- Species range shifts towards higher latitudes or altitudes

- Disruption of ecosystems due to changes in habitat conditions

- Extinctions as a result of inability to adapt quickly

5. Economic Impacts

The economic consequences of climate change are multifaceted and disruptive, affecting various sectors:

- Reduced agricultural output, impacting food prices and global trade

- Increased costs for water management infrastructure

- Loss of income for industries tied to ecosystem services, such as tourism or fishing

- Greater investment needed in climate-resilient infrastructure

- Economic losses from extreme weather events

6. Social Impacts

Climate change exacerbates social inequalities by disproportionately affecting already vulnerable populations such as the poor, elderly, or those living in low-latitude regions. Some societal implications include:

- Increased migration and displacement due to food scarcity, sea level rise, or conflict

- Strained public health systems and infrastructure

- Greater inequality between more and less developed regions

- Potential for social unrest due to resource scarcity

The wide-ranging impacts of climate change and rising temperatures demand a comprehensive understanding and adaptive approach to minimize potential threats. This chapter has outlined the primary consequences that societies will face and highlighted the need for urgent action through mitigation strategies, adaptation planning, and international cooperation to build a more resilient and sustainable world.

Let's examine the causes and consequences of rising temperatures, as well as some of the unique challenges that these countries face. Furthermore, we will explore the response and adaptation strategies of these nations to mitigate the effects of climate change on their economies, ecosystems, and communities.

1. Overview of Climate Change and Rising Temperatures:

Before we focus on specific countries, it is essential to understand the underlying reasons for climate change and its relationship with increasing temperatures. This section will cover the science behind global warming, including the greenhouse effect and anthropogenic factors contributing to this phenomenon.

2. Effects of Rising Temperatures in Southern Hemisphere Countries:

This section will provide an overview of the myriad ways in which rising temperatures due to climate change affect southern hemisphere countries. It will touch upon topics such as:

a) Heatwaves

b) Droughts

c) Impacts on agriculture

d) Disruption of ecosystems

e) Sea-level rise

3. Case Studies in the Southern Hemisphere:

This section will offer an in-depth look at how specific nations cope with temperature spikes brought on by climate change. The selected countries serve as representatives of different regions in the southern hemisphere.

a) Australia: Confronting Wildfires and Coral Reef Bleaching

b) Argentina: Shifts in Agriculture and Livestock Production

c) South Africa: Droughts Stressing Water Scarcity

d) Brazil: Deforestation Worsening Climate Change

4. Adaptation and Mitigation Strategies:

Having explored some illustrative cases, this section delves into current strategies employed by nations from different continents, seeking both adaptation and mitigation responses.

a) Green Energy Investments

b) Sustainable Agriculture Practices

c) Water Management Systems

d) Coastal Infrastructure Improvements

5. International Cooperation and Support:

The chapter will conclude by exploring the importance of global cooperation in addressing a challenge that transcends political boundaries. It will encompass the role of international organizations and treaties, such as the Paris Agreement and United Nations Framework Convention on Climate Change (UNFCCC). Furthermore, it will analyze how developed nations can assist southern hemisphere counterparts in coping with these pressing challenges.

By the end of this chapter, readers will have a comprehensive understanding of the multifaceted impacts of rising temperatures in different countries of the southern hemisphere due to climate change. The provided analysis and case studies aim to offer insights into both challenges faced and solutions adapted as these nations strive towards a more climate-resilient future.

Climate Change Temperature Targets and the Consequences of Exceeding Them

THIS SECTION FOCUSES on climate change temperature targets, the consequences of exceeding these targets, and the tipping points that can be triggered if target temperatures are breached.

Section 1: Climate Change Temperature Targets

1.1 International Agreements on Temperature Targets

The most significant international agreement that sets temperature targets is the Paris Agreement, ratified in 2016 by 189 countries. The central aim of this accord is to strengthen the global response to the threat of climate change by keeping a global temperature rise this century well below 2 degrees Celsius above pre-industrial levels and pursuing efforts to limit the temperature increase even further to 1.5 degrees Celsius.

1.2 Scientific Basis for Temperature Targets

These temperature targets are grounded in scientific evidence, particularly from the Intergovernmental Panel on Climate Change (IPCC). The rationale is twofold:

a) Limiting warming to 1.5 degrees Celsius would reduce many climate change-related risks compared with a 2-degree Celsius increase.

b) The risk of crossing critical thresholds or tipping points, which can lead to abrupt and irreversible changes in Earth's systems, increases significantly above 2 degrees Celsius.

Section 2: Consequences of Exceeding Temperature Targets

Failure to meet climate change temperature targets has profound consequences, ranging from intensified environmental impacts to increased social and economic burdens.

2.1 Environmental Consequences

a) Increased frequency and severity of extreme weather events such as heatwaves, floods, droughts, and tropical storms.

b) Accelerated melting of polar ice sheets, contributing to increased sea level rise and flooding of coastal regions.

c) Disruption of ecological systems, leading to species loss and extinctions, as well as imbalances in ecosystems.

d) Ocean temperature rise, acidification, and deoxygenation, leading to the decline of marine biodiversity and disruption of global fisheries.

2.2 Social and Economic Consequences

e) Increased water scarcity, leading to conflicts and mass migration.

f) Reduced agricultural yields and increased food insecurity.

g) Negative impacts on public health, with greater exposure to heat stress, disease transmission, and poor air quality.

h) Increased financial strain due to the cost of disaster response and adaptation measures.

Section 3: Tipping Points in Climate Change

Tipping points represent critical thresholds in Earth's systems that, once crossed, can lead to abrupt and often irreversible changes. Temperature targets are related to these tipping points, as breaching the targets heightens the risk of passing them.

3.1 Key Tipping Points

a) Collapse of polar ice sheets: this can occur due to melting from warmer temperatures or destabilization from ocean warming. Loss of ice significantly contributes to sea level rise.

b) Disruption of ocean circulation patterns: Climate change affects the North Atlantic circulation patterns that contribute significantly to global climate stability. This can have cascading effects on both regional and global scales.

c) Large-scale release of methane: Significant amounts of methane hydrates beneath permafrost or ocean sediments could be released if temperatures continue to rise, further amplifying global warming.

d) Die-off or collapse of Amazon rainforest: The Amazon serves as a major carbon sink; its destruction could weaken the planet's capacity for carbon absorption.

Adhering to climate change temperature targets is essential for reducing adverse environmental impacts as well as social and economic burdens. Moreover, keeping below target temperatures. Let's look at these tipping points in more detail.

Collapse of polar ice sheets: this can occur due to melting from warmer temperatures or destabilization from ocean warming. Loss of ice significantly contributes to sea level rise.

One particularly concerning climate change tipping point is the collapse of polar ice sheets. In this chapter, we will explore how warmer temperatures and ocean warming contribute to ice sheet destabilization, the potential consequences of such collapse on global sea levels, and the implications of reaching this tipping point.

1. Understanding Polar Ice Sheets

1.1 Formation and Composition

Polar ice sheets are vast expanses of glacial land ice that cover Greenland and Antarctica. Formed from snow accumulating over thousands of years and compacting into ice, they hold close to 99% of the world's freshwater reserves.

1.2 Role in Global Climate Regulation

Ice sheets play an essential role in regulating Earth's climate by reflecting solar radiation back into space and acting as a temperature buffer by storing immense amounts of heat.

2. The Dangers of Melting Ice Sheets

2.1 Temperature Increase

The Intergovernmental Panel on Climate Change (IPCC) anticipates a significant increase in average global temperatures this century, accelerating ice sheet melting through various processes.

2.2 Destabilization from Ocean Warming

Ocean warming is another contributing factor to the destabilization of polar ice sheets as it causes their edges to melt even faster than surface melting alone.

---

3. *SEA LEVEL RISE FROM Ice Sheet Collapse*

3.1 The Contribution to Sea Level Rise

Melting polar ice sheets can have dire consequences for rising sea levels as they hold enough freshwater to raise global mean sea level by around 65 meters if completely melted.

3.2 Impact on Coastal Communities

The increased rate of sea-level rise adversely affects coastal communities across the world, causing flooding, erosion, and loss of habitable land and infrastructure.

4. Beyond the Tipping Point

4.1 Irreversible Changes

When the climate reaches a tipping point, such as that posed by the collapse of polar ice sheets, there might be no going back, leading to a series of irreversible changes in Earth's climate system.

4.2 A Domino Effect

The collapse of polar ice sheets could trigger other tipping points, such as permafrost melting that releases additional greenhouse gases or ocean circulation changes that further intensify climate change.

5. *Adaptation and Mitigation Strategies*

5.1 Reducing Greenhouse Gas Emissions

In order to slow down or prevent the collapse of polar ice sheets, efforts to reduce greenhouse gas emissions on a global scale are crucial.

5.2 Preparing for Sea Level Rise

Governments and communities must develop strategies to manage and adapt to sea levels rise, including rebuilding infrastructure and developing better warning systems for coastal areas.

THE COLLAPSE OF POLAR ice sheets is a potentially devastating tipping point in our warming world. By understanding its causes, effects, and interconnections within the climate system, we can help mitigate its impacts and work towards a more resilient future. The challenges posed by collapsing ice sheets serve as another reminder of the urgent need for international cooperation on reducing greenhouse gas emissions and preparing for the consequences of a changing climate.

Disruption of ocean circulation patterns: Climate change affects the North Atlantic circulation patterns that contribute significantly to global climate stability. This can have cascading effects on both regional and global scales.

Ocean circulation patterns play a crucial role in maintaining the global climate by redistributing heat and influencing the exchange of carbon dioxide between the atmosphere and ocean. One of the key systems is the Atlantic Meridional Overturning Circulation (AMOC), which consists of a complex system of currents spanning the North Atlantic. Climate change has triggered perturbations in these circulation systems, which could lead to significant and potentially irreversible consequences for the planet. These next few paragraphs will discuss the importance of ocean circulation patterns, how climate change affects these systems, and the cascading impacts on regional and global scales.

Understanding Ocean Circulation Patterns

The AMOC is responsible for circulating warm surface waters from the tropics towards high latitudes while transporting cold, deep waters from polar regions towards the equator. This complex cycle helps balance Earth's temperature by redistributing heat around the globe. The Gulf Stream, a vital component of AMOC, moderates temperatures in Western Europe by ushering warm tropical waters into the region.

Without this current, those regions would experience more extreme fluctuations between summer and winter temperatures.

Effect of Climate Change on Ocean Circulation Patterns

Human-induced climate change has resulted in rapid increases in atmospheric temperature and unprecedented melting rates of glaciers and ice sheets, causing freshwater input into oceans to rise dramatically. This sudden increase in freshwater interferes with ocean currents' natural course by altering water salinity levels coursing through different regions. Consequently, this alters the water density and disrupts temperature exchange mechanisms—the driving forces behind ocean currents.

Cascading Effects on Regional Scales

The deterioration of ocean circulation patterns has numerous impacts on regional scales. Western Europe's stability may be threatened as weakened Gulf Stream currents lead to increasingly extreme weather events—disruptive to societies that have developed around more temperate conditions. Moreover, altered oceanic pathways impact the migration and distribution of marine species, which could potentially disrupt ecosystems and local fishing industries.

Even more significant, slower AMOC might exacerbate regional sea-level rise along the U.S. East Coast due to disrupted deepwater formation processes. This effect poses severe risks to coastal cities such as New York and Miami, with potentially catastrophic consequences for housing, infrastructure, and human lives.

Cascading Effects on Global Scales

On a global scale, should the AMOC collapse completely, it would have drastic implications for the planet's climate. Areas once warmed by ocean circulation could experience sudden cooling. Moreover, this collapse might trigger other tipping points related to ice sheets and sea level rises or even influence tropical monsoon patterns by changing ocean-atmosphere interactions.

Furthermore, changes in ocean currents impact the sequestration of greenhouse gases like carbon dioxide when cooler, carbon-rich surface waters sink into deeper layers of the ocean. A disruption in this process could significantly alter the balance of these gases in the atmosphere—and consequently accelerate climate change even more.

The disruption of ocean circulation patterns presents a unique threat as it acts as both an indicator and accelerator of climate change. Understanding this tipping point is critical to mitigating its regional and global impacts. By advocating for carbon emission reduction policies and fostering international cooperation around climate science research, society can work to preserve vital ocean currents like the AMOC—preventing a cascade effect on Earth's delicate climate system.

Large-scale release of methane: Significant amounts of methane hydrates beneath permafrost or ocean sediments could be released if temperatures continue to rise, further amplifying global warming.

The threat of climate change has become increasingly dire, with countless studies demonstrating its devastating impacts on the planet. One particular tipping point that has garnered significant attention from both scientists and policymakers is the large-scale release of methane from beneath permafrost or ocean sediments due to rising temperatures.

In this chapter, we will delve into the intricacies of methane release, its impacts on global warming, and the broader implications for our environment and society.

I. Understanding Methane Hydrates

To grasp the potential implications of a large-scale methane release, it is essential to understand the nature of methane hydrates. Methane hydrates are solid, ice-like structures that form when methane molecules become trapped within a lattice of water molecules. These hydrates can be found in significant quantities in areas with low temperatures and high pressures, such as under permafrost or deep ocean sediments.

II. How Rising Temperatures Trigger Methane Release

As global temperatures continue to rise due to human-induced greenhouse gas emissions, there is an increasing risk that these methane hydrates will destabilize and release their stored methane in large quantities. This could happen through:

A. Thawing Permafrost: Permafrost acts as a natural cap that keeps methane hydrates sealed within the ground. As temperatures rise, permafrost melts and loses its strength, potentially allowing for the escape of trapped methane.

B. Ocean Warming: Warming ocean temperatures can trigger methane releases from deep-sea sediments by prompting the destabilization of hydrate formations or chemically disassociating them.

III. Methane: A Potent Amplifier of Global Warming

The potential large-scale release of methane is concerning because it is a powerful greenhouse gas with a global warming potential approximately 28-36 times greater than carbon dioxide over a 100-year period. Methane's release into the atmosphere can accelerate climate change, leading to even more extreme weather events, ice melts, sea level rise, and a range of other negative impacts.

IV. TIPPING POINT IMPLICATIONS

The large-scale release of methane represents a tipping point in the climate change narrative because once initiated, it disrupts the planet's climate system, rapidly accelerating the escalation of global warming. This could lead to:

A. Accelerated Ice Melt: Methane-driven temperature increases can accelerate ice cap and glacier melt rates, causing rapid sea level rise.

B. Compound Feedback Loops: The effect of methane release can worsen existing feedback loops, such as the loss of Arctic ice resulting in greater heat absorption.

C. Biodiversity Loss: Ecosystems and species unable to adapt quickly enough could face extinction due to rapid environmental shifts.

V. MITIGATION AND ADAPTATION Strategies

Finally, we will explore possible mitigation and adaptation strategies to address the threat posed by large-scale methane release:

A. Methane Leak Detection and Repair: Implementing better technologies for monitoring and fixing leakages from fossil fuel infrastructure can minimize emissions.

B. Carbon Sequestration Techniques: Developing methods for capturing and storing carbon dioxide from power plants and other industrial sources could help reduce atmospheric warming that contributes to hydrate melting.

C. Adapting Infrastructure and Coastal Planning: Planning for rising sea levels should be integrated into long-term strategies for coastal communities to minimize disruption.

As our understanding of climate change tipping points evolves, understanding the significance of large-scale methane releases is crucial in designing effective mitigation and adaptation policies. By recognizing potential impacts of destabilizing methane hydrates on the earth's climate system, we can work together towards safeguarding our planet's future.

Die-off or collapse of Amazon rainforest: The Amazon serves as a major carbon sink; its destruction could weaken the planet's capacity for carbon absorption.

The Amazon rainforest plays a crucial role in regulating the global climate, acting as both a carbon sink and an air purifier. Spanning over 6.7 million square kilometers across nine South American countries, it is home to about 400 billion individual trees and a diverse array of flora and fauna. As climate change accelerates, one pressing concern is the potential collapse of the Amazon rainforest - a major tipping point that could weaken the planet's capacity for carbon absorption and push our ecosystems beyond their limits.

1.2 CARBON STORAGE and Sequestration in the Amazon

Forests serve as carbon sinks by absorbing large quantities of atmospheric carbon dioxide during photosynthesis and storing it in their biomass. The Amazon rainforest alone holds approximately 90-140 billion metric tons of carbon, representing nearly 10 years' worth of global emissions from human activities. However, human-induced factors such as logging and forest fires contribute to deforestation and degradation, reducing its ability to sequester carbon.

1.3 Drivers of Deforestation in the Amazon

The primary drivers of deforestation in the Amazon are agriculture, logging, infrastructure development, and mining. Cattle ranching accounts for over two-thirds of deforestation, followed by crop cultivation like soybean farming. Illegal logging further exacerbates this issue by disrupting ecosystems, releasing stored carbon back into the atmosphere.

1.4 THE TIPPING POINT Scenario: Collapse of the Amazon Rainforest

As we increase greenhouse gas emissions from human activities, alter precipitation patterns due to climate change, and disturb the delicate balance within the ecosystem through deforestation, the Amazon rainforest edges closer to its tipping point - a die-off or complete collapse.

Scientists estimate that if 20-25% of the forest is lost or degraded, reduced rainfall could trigger a large-scale die-off, transforming the Amazon from a luscious rainforest into a savanna-like ecosystem. This would severely impact its ability to store and absorb carbon, exacerbating global warming.

1.5 Consequences of the Amazon Rainforest Collapse

If the Amazon were to collapse, the global implications would be devastating:

1. LOSS OF BIODIVERSITY: As the Amazon hosts one in ten known species on Earth, its collapse would lead to a loss of irreplaceable biodiversity and global ecological imbalances.

2. Droughts and floods: The Amazon's water cycle affects rainfall patterns far beyond its borders. Its collapse could trigger more frequent and severe droughts and floods worldwide.

3. Impact on indigenous populations: The collapse would endanger the livelihoods, culture, and survival of millions of indigenous people relying on the forest for sustenance and medicines.

1.6 Preventing the Tipping Point: Mitigation Strategies

To protect the Amazon rainforest and avoid reaching this tipping point, concerted efforts from international communities, governments, businesses, and individuals are needed:

1. Establish effective forest conservation policies and sustainable land-use management.

2. Strengthen law enforcement against illegal logging and promote certified timber production.

3. Encourage sustainable agriculture practices that reduce pressure on forests.

4. International collaboration to support economic incentives and funding for reforestation projects.

The potential collapse of the Amazon rainforest poses dire consequences for global ecosystems, climate stability, and humanity at large. By understanding the gravity of this climate change tipping point and taking necessary actions to mitigate deforestation, we can preserve this invaluable natural resource for future generations while combating climate change simultaneously.

Chapter Eight
Increased frequency and severity of natural disasters

———

The Escalating Threat of Natural Disasters due to Climate Change Let's explore various consequences that could stem from these disasters and examine some recent incidents directly linked to climate change.

Understanding the potential risks is essential for communities, policymakers, and emergency management agencies to adequately prepare for and address these challenges.

Section 1: Consequences of Increased Frequency and Severity of Natural Disasters

1.1 Greater loss of life: As natural disasters escalate in frequency and intensity, fatalities are expected to rise, overwhelming emergency response services.

1.2 Economic impacts: Severe disasters result in substantial economic strain caused by infrastructure damage, disruption of supply chains, and lost productivity.

1.3 Displacement of populations: An increase in catastrophic events could lead to extensive dislocation and migration due to inaccessible or uninhabitable areas.

1.4 Food insecurity: The growing prevalence of natural disasters

1.5 Mental health consequences: The emotional toll from frequent catastrophic events can lead to a host of mental health disorders like anxiety, depression, and post-traumatic stress disorder (PTSD).

Section 2: Recent Natural Disasters Attributed to Climate Change

2.1 Hurricane Iota (2020): This Category 5 hurricane devastated parts of Central America with widespread flooding and violent winds, resulting in massive destruction and loss of life. The warming climate is believed to have contributed to its intensity.

2.2 Wildfires in California (2020): The devastating wildfires across California were exacerbated by extremely dry conditions due to prolonged droughts associated with climate change. This catastrophe resulted in loss of life, displacement of communities, and billions of dollars in damages.

2.3 Super Cyclone Amphan (2020): Striking India and Bangladesh in May 2020, Amphan's devastating impact was linked to unusually high sea surface temperatures caused by climate change.

2.4 Australian Bushfires (2019-2020): The enormous bushfire crisis in Australia was intensified by factors linked to global warming, such as droughts, extreme heat, and shifting weather patterns. These fires significantly impacted the environment, wildlife, and local communities.

2.5 HEATWAVES IN EUROPE (2018-2019): Increasingly intense heatwaves in Europe have been attributed to a warming climate and are causing severe impacts on public health, agriculture, and infrastructure.

The increasing severity and frequency of natural disasters due to climate change present significant challenges for the safety and well-being of global communities. By understanding these potential outcomes and analyzing recent events, society becomes better equipped to deal with such disasters as we work to reverse the harmful impacts of climate change.

Let's look a bit closer at Displacement of populations: An increase in catastrophic events could lead to extensive dislocation and migration due to inaccessible or uninhabitable areas.

A growing body of evidence suggests that climate change is leading to an increase in the frequency and intensity of natural disasters around the world. The implications of these changes are far-reaching, with considerable impacts on ecosystems, infrastructure, and human populations. One of the most significant results of this turbulent shift is the displacement of people as they search for more stable living environments. This chapter delves into the complex issue of climate-induced migration, with a particular focus on population movements spurred by natural disasters and conflicts exacerbated by climate change.

The Destructive Power of Natural Disasters

In recent decades, natural disasters have become more frequent and more deadly due to climate change. The devastation wrought by these events can be immediate and lasting, rendering vast swaths of land uninhabitable and forcing residents to a desperate search for viable alternatives. From hurricanes to floods to wildfires, the planet's changing climate has unleashed disastrous consequences that stand to displace millions in the coming years.

THE MIGRATION IMPERATIVE

As populations find themselves in increasingly vulnerable environments due to events precipitated by climate change, migration becomes less a choice and more a necessity for survival. The potential number of people who could be displaced from their homes and forced to migrate due to catastrophic events is staggering. Current predictions indicate tens or even hundreds of millions may need to move northward in response to a combination of worsening conditions at home and more attractive prospects elsewhere.

It is crucial to recognize that climate change does not operate as an isolated force; rather, it has profound reverberations across political landscapes. As resources such as water become scarcer and competition for them intensifies, struggles over these critical lifelines have exacerbated pre-existing tensions between communities and fueled armed conflicts around the world. These wars destabilize regions further, prompting even more people to seek refuge in safer territories.

Towards a More Resilient Future

Addressing the issue of climate-induced displacement requires innovative and comprehensive solutions to tackle both its proximate and underlying causes. To that end, multiple strategies must be implemented concurrently, including bolstering the resilience of at-risk communities, providing support for those already displaced, and investing in concerted global efforts to curb greenhouse gas emissions. This requires collaboration between governments, industries, and civil society towards unified action in combating this existential threat.

As the world faces an unprecedented increase in natural disasters and conflicts spurred by climate change, climate-induced migration is becoming an increasingly visible aspect of human adaptation to shifting circumstances. By understanding the complex interplay between these factors and developing targeted interventions, we can work together to protect vulnerable populations and forge a more resilient, sustainable path into the future.

Chapter Nine
Sea level rise and loss of habitable land

As the global climate continues to change, one of the most significant consequences is the rise in sea levels. This phenomenon, driven by factors such as melting ice sheets and thermal expansion of the ocean, poses a considerable threat to human societies and ecosystems worldwide. As we witness varying degrees of temperature rise due to climate change, understanding the implications for habitable land becomes crucial. In this chapter, we delve into the consequences of sea-level rise at different temperature projections, identify the countries and cities most at risk, and estimate the amount of land that could potentially be lost.

Temperature Rise Levels and Sea Level Projections

To analyze the impact of sea level rise on habitable land, it is essential to correlate increasing temperatures with projected sea levels. We will consider three primary temperature rise scenarios:

1. Low-end scenario (1.5°C increase by 2100)

2. Intermediate scenario (2°C increase by 2100)

3. High-end scenario (4°C increase by 2100)

For each of these scenarios, we will outline the anticipated changes in sea levels and associated consequences for habitable regions.

Countries and Cities Most at Risk

Given varying topography, infrastructure capabilities, and population densities, certain countries and cities are more susceptible to sea-level rise than others. The nations most at risk include:

1. Bangladesh

2. Vietnam

3. Indonesia

4. India

5. China

Within these nations lie major cities where millions reside in close proximity to coastal areas such as Dhaka (Bangladesh), Ho Chi Minh City (Vietnam), Jakarta (Indonesia), Kolkata (India), and Shanghai (China). As such, these urban centers are highly susceptible to significant losses of land.

In addition to these at-risk nations, many small island nations and territories, such as the Maldives, Marshall Islands, and the Solomon Islands, are severely endangered by rising sea levels.

Land Loss Projections

To detail potential land losses caused by sea level rise, we will use a combination of existing research, models, and projections associated with each temperature rise scenario. The estimations will result in a range of potential land losses for both global impact and specific high-risk countries and cities. The figures will display the severity of the situation for different temperature scenarios, emphasizing the urgency to combat climate change across our interconnected world.

As we grapple with the unprecedented challenges posed by sea-level rise due to climate change, understanding potential land losses at various temperature scenarios is crucial. By identifying high-risk countries and cities and analyzing predictions related to coastal areas' inundation, we amplify our awareness on the necessity for immediate action in mitigating climate change and protecting vulnerable regions worldwide. This chapter represents merely a stepping stone toward understanding the extent of transformations unfolding before us, highlighting the importance of collaboration amongst nations to adapt to and mitigate the consequences of a changing planet.

Chapter Ten
Effects on agriculture, water supplies, and food security

1. Droughts and water scarcity leading to crop failures

    1. Agriculture-dependent economies.
    2. Disruption in food production and distribution networks

CLIMATE CHANGE IS A global phenomenon that has garnered significant attention in recent decades, as the implications of a warming planet become more apparent. One of the most critical areas affected by climate change is agriculture, which not only has ramifications for water supplies but also impacts food security worldwide. This chapter delves into the intricate relationship between climate change and its effects on agriculture, water resources, and food security by examining three key aspects: droughts and water scarcity leading to crop failures; agriculture-dependent economies; and the disruption of food production and distribution networks.

Section 1: Droughts and Water Scarcity Leading to Crop Failures

Droughts have long been a menace to agriculture as they result in reduced water availability for crops, leading to diminished yields or even complete crop failure. Climate change has exacerbated this issue, as increasing average global temperatures intensify evaporation rates and shift precipitation patterns, resulting in more frequent and severe droughts.

Consequently, reduced access to water translates into significant challenges for farmers who must then contend with declining crop productivity.

Moreover, water scarcity does not only stem from drought conditions but can also be exacerbated by unsustainable irrigation practices or the over-extraction of groundwater. Climate change further complicates matters by introducing uncertainties in weather patterns and precipitations forecasts. This section will delve into the causes of droughts and water scarcity, their direct impacts on agricultural productivity, and potential adaptation strategies for mitigating these effects.

Section 2: Agriculture-Dependent Economies

Many countries around the world have economies heavily reliant on agriculture. Oftentimes, these nations have large rural populations whose livelihoods depend on farming and access to arable land.

HOWEVER, CLIMATE CHANGE poses an acute threat to such agriculture-dependent economies due to increased risks of extreme weather events like floods or heatwaves that can detrimentally affect crop yields.

This section will explore how changes in agricultural productivity due to climate change-related stressors can threaten the stability of agriculture-dependent economies. Furthermore, it will consider how economic diversification and support initiatives may contribute to enhanced resilience in the face of climate change-related agricultural challenges.

Section 3: Disruption in Food Production and Distribution Networks

Climate change not only affects crop growth but also has far-reaching implications for food production and distribution networks. For instance, shifts in global weather patterns and extreme weather events can result in transportation disruptions affecting the movement of food products from their point of origin to markets.

In addition, further complexities arise when considering issues such as food waste during transportation or susceptibility to pests and disease as climate-induced changes provide new risks that farmers and distributors must contend with. This section investigates how climate change disrupts various aspects of food production and distribution networks and discusses potential solutions for building more resilient systems that can better cope with these challenges.

Climate change presents numerous challenges to agriculture, water resources, and food security, impacting not only crop productivity, but also the wellbeing of millions of people worldwide. Addressing these complex issues requires a multifaceted approach, integrating sustainable practices alongside innovative strategies that bolster resilience throughout every stage of the agricultural process. Through this understanding, it becomes imperative for governments, policy-makers, farmers, and consumers alike to join forces in mitigating the negative impacts of climate change on agriculture and food systems worldwide.

Chapter Eleven

Migration Patterns and Potential Exodus Due to Climate Change

Climate change has become a critical global issue with significant implications on the environment and human societies. Let's discuss the possible migration patterns and potential exodus of populations if nothing is done to mitigate climate change. We will focus on destinations such as Europe and North America, analyzing their attractiveness to climate migrants, and discussing the social, economic, and political ramifications that may arise.

The Driving Forces behind Climate Migration

Climate change impacts can manifest in various ways such as increase in average temperatures, more frequent extreme weather events, rising sea levels, and changes in precipitation patterns. These impacts can lead to loss of arable lands, water scarcity, food insecurity, heightened risk of diseases, and loss of livelihoods – all factors contributing to forced migration.

Destinations for Climate Migrants

1. Europe

There are several reasons why Europe may emerge as a popular destination for climate migrants:

- Proximity: Certain European countries are geographically close to nations facing severe climate challenges such as those in North Africa and the Middle East.

- Economic opportunities: Europe has a relatively robust economy which makes it attractive for migrants seeking better living conditions.

- Social programs: European countries offer reliable social services such as healthcare, education, and housing support that can help create new opportunities for advancement.

2. NORTH AMERICA

North America also presents itself as a potential destination for climate migrants:

- Capacity to accommodate: With its vast land area, there is room for population growth in this region.

- Economic prospects: The United States and Canada have strong economies that can offer better opportunities for migrants in terms of jobs and quality of life.

- Existing diaspora communities: For people displaced by climate change, having an established community from their home country can ease the process of integration into a new society.

Projected Migration Patterns

According to the International Organization for Migration, estimates suggest that there could be as many as 200 million people displaced due to climate change by 2050. Based on current trends, the migration patterns can be broadly categorized as:

- Intra-regional migration: People from the most affected countries will initially move to neighboring countries within their region.

- Inter-regional migration: As conditions worsen, more distant destinations such as Europe and North America will become increasingly appealing.

Challenges and Opportunities

The potential exodus of climate migrants can create significant challenges, including:

- Pressure on resources: A sudden influx of migrants may strain public services, health care systems, and infrastructure in destination countries.

- Social integration: Large-scale migration can lead to social tensions if there is a perceived competition for jobs or cultural differences between the local population and new arrivals.

- Political backlash: Anti-immigrant sentiments may lead to political instability in some regions.

Conversely, this phenomenon can also provide certain opportunities such as:

- Labor force replenishment: Aging populations in Europe and North America can benefit from an influx of working-age migrants.

- Economic benefits: Migrants often contribute to economic growth through their labor and consumption.

- Greater awareness: This situation can highlight the urgency of addressing climate change with more concerted global efforts.

A potential exodus of people due to climate change highlights the need for immediate action on multiple fronts. Our collective efforts should involve not only addressing the root causes of climate change but also enhancing global collaboration to manage large-scale migration. By understanding these migration patterns and possible repercussions, we can better prepare for a future defined by altered social landscapes and environmental realities

Chapter Twelve

——

Case studies of countries experiencing major shifts in population due to climate conditions including Bangladesh's sea level rise and flooding South Africa. The desertification and persistent droughts around the world

As adverse climate conditions like sea level rise, flooding, desertification, and persistent droughts continue to intensify, millions of people are forced to migrate in search of better lives and livelihoods. In this section we explore various case studies from around the world to understand the extent and impact of these shifts on affected countries.

Case Study 1: Bangladesh - Sea Level Rise and Flooding

Bangladesh is one of the most climate-vulnerable countries in the world due to its low-lying landscape and exposure to extreme weather events. Rising sea levels have caused widespread flooding in the coastal areas, leading to the displacement of millions of people.

A. Causes and extent of sea level rise in Bangladesh

B. Impact on agriculture, livelihoods, and infrastructure

C. Government initiatives, adaptation measures, and international support

D. The future outlook for Bangladesh's population shifts

Case Study 2: South Africa - Desertification and Persistent Droughts

South Africa is undergoing a shift caused by the desiccation of large parts of its territory as well as long-lasting droughts which are affecting water availability for both agriculture and drinking purposes.

A. Factors contributing to desertification in South Africa

B. Impact on water resources, agriculture, food security, and economy

C. Government policies and local communities' efforts in combating desertification

D. Future scenarios for populations affected by desertification in South Africa

Case Study 3: Other Countries Facing Climate-Induced Displacements

Beyond our focused case studies on Bangladesh and South Africa, numerous other countries are dealing with similar challenges caused by climate change.

A. Pacific Island nations experiencing rising sea levels (Kiribati, Tuvalu)

B. Drought-stricken regions forcing migration (Horn of Africa, Central America)

C. The complex intersection of climate change, conflict, and migration (Syria)

It is evident that climate change-induced population shifts are not isolated instances but a global phenomenon. Countries affected by these migrations are not only challenged by the direct impacts of environmental degradation, but also by socio-economic consequences. Collaborative efforts on national and international levels are required to mitigate the damage to affected communities and curb further progression of these detrimental climatic conditions.

Chapter Thirteen

Economic Consequences for Europe and North America

For example: Labour market changes. Public services capacity strain. Social adjustment challenges. Cultural integration, discrimination

Climate change and economic migration are two main global challenges confronting Europe and North America. Over the years, these phenomena have vastly altered the economic landscape for these continents, forcing governments, businesses, and individuals alike to adapt to new realities. This essay examines the multiple economic consequences experienced by Europe and North America as a result of these transformations. Focusing on labor market changes, public service capacity strain, social adjustment challenges and cultural integration considerations, this essay calls for concerted efforts to address these issues for a sustainable future.

I. Labor Market Changes

1. Shifts in labor demand patterns: Climate change has led to significant alterations in industry structures across Europe and North America. Sectors such as agriculture, manufacturing, tourism and infrastructure development have faced unique challenges that require workforce re-alignment and reskilling.

2. Labor mobility: Economic migration has led to an influx of workers from different countries seeking contracts or permanent jobs. This increased flow creates both opportunities and problems. On one hand, it can help fill critical gaps in the labor market; on the other hand, it can cause competition among locals for available jobs.

3. Skills mismatch: The continuous changes in labor markets strain educational institutions that struggle to keep up with the trends requiring curricula updates to better prepare students for positions in evolving industries.

II. Public Services Capacity Strain

1. Healthcare sector: The influx of economic migrants often puts pressure on national healthcare systems. Migrants' health needs could impose significant additional costs onto governments that need to ensure available resources match growing demands.

2. Education system: Schools and institutions face challenges in integrating migrants into their classrooms while retaining high-quality education standards.

3. Housing sector: Cities face strain as they develop social housing plans for growing populations and the infrastructures to go along with it.

III. Social Adjustment Challenges

1. Pressure on civic resources: The inflow of economic migrants puts tremendous pressure on existing infrastructure such as transportation, sanitation, and other civic resources.

2. Economic disparities: Inequality arises when locals perceive incomers as taking opportunities from pre-existing labor forces, creating tension between communities.

3. Political tensions: Increased competition for resources and perceived cultural threats can lead to political tensions revolving around immigration policies.

IV. CULTURAL INTEGRATION and Discrimination

1. Cultural exchange: The influx of economic migrants presents an opportunity for cross-cultural exchange, enriching societies.

2. Adaptation strategies: Both economic migrants and host societies have to be deliberate in fostering a sense of unity by facilitating language learning and encouraging social assimilation.

3. Discrimination and xenophobia: The difficulties assimilating may result in higher instances of discrimination and xenophobia against migrants in the workplace, educational establishments, and other public areas.

Europe and North America face complex economic consequences resulting from climate change and economic migration. Addressing these multifaceted challenges requires a multidirectional approach, focusing on human capital investments, ensuring cohesive social integration, embracing cultural diversity, and developing supportive policies that facilitate adaptability to changing global conditions.

Chapter Fourteen
Climate Change Wars: Resource Conflicts

Competition for scarce resources such as water, arable land, energy sources

Examples of conflicts arising from climatic factors (Darfur, Syria)

Regional disputes over common resources (water rights between nations sharing rivers)

As the world grapples with the interconnected challenges of climate change and resource scarcity, it becomes increasingly apparent that these issues may have profound implications for global peace and stability. In this section we suggest the potential for future conflicts arising from climate change and resource scarcity, taking into account geopolitical trends, societal pressures, and technological advancements.

Section 1: Climate Change as a Threat Multiplier

1.1. The broad-ranging impacts of climate change - from rising sea levels to extreme weather events - pose significant threats to human societies around the globe.

1.2. The exacerbation of existing tensions between nations, as climate-induced migration leads to competition over basic resources such as land, water, and energy.

1.3. The potential for non-state actors to exploit climate-related vulnerabilities, contributing to radicalization and the rise of eco-terrorism.

Section 2: Geopolitical Flashpoints in a Resource-Scarce World

2.1. Water scarcity as a source of tension in transboundary river basins (e.g., Nile, Tigris-Euphrates, Mekong), with increasing demand for water leading to political disputes and even armed confrontations.

2.2. Competition over dwindling energy resources, particularly in regions with vast fossil fuel reserves (e.g., Middle East) or emerging renewable energy technologies (e.g., rare earth minerals).

2.3. Territorial disputes exacerbated by diminishing landmass due to rising sea levels, particularly in low-lying island nations and coastal regions.

Section 3: Socioeconomic Dimensions of Climate Change Wars and Resource Conflicts

3.1. The role of climate-induced migration in exacerbating social tension and xenophobia within destination countries.

3.2. Economic disparities arising from unequal distribution of resources amid increased scarcity, further perpetuating conflict, and instability.

3.3. The failure of state institutions to effectively address climate and resource-related challenges, leading to internal turmoil and political unrest.

Section 4: Technological Solutions and Strategic Adaptation

4.1. The development of innovative technologies for mitigating climate change and its effects, such as geoengineering, carbon capture, and renewable energy systems.

4.2. Fostering international cooperation and diplomacy in the realms of resource sharing, environmental management, and conflict prevention.

4.3. Encouraging proactive measures to build societal resilience against resource scarcity through programs for sustainable agriculture, infrastructure development, and social inclusivity.

The potential for climate change wars and resource conflicts in the future is a daunting prospect that requires urgent attention from policymakers, technologists, and civil society alike. By anticipating these challenges ahead of time and fostering collaborative approaches to problem-solving, we can strive to mitigate the profound impacts of climate change on human societies while promoting peace and global stability in an increasingly resource-scarce world.

Climate change is increasingly becoming a global concern, with numerous consequences like fluctuating temperatures, natural disasters, and threat to biodiversity. One lesser-discussed but significant effect is the upsurge in resource conflicts and wars. In regions like Darfur, a region located in the western part of Sudan, climate change has aggravated existing socio-political tensions and sparked new battles over scarce resources such as water and fertile land. This essay will discuss the future of climate change wars and resource conflicts in Darfur, highlighting how environmental factors exacerbate societal divisions and the potential solution for mitigating these challenges.

The Roots of Conflict:

To understand the emergence of climate change conflicts and resource wars in Darfur, it is crucial to recognize its historical context. Darfur has long struggled with ethnic tension between Arab herders (predominantly nomadic) and African farmers (settled communities). The region's social fabric is continuously strained due to competing interests for land, water, and grazing grounds which roots back to inequality ingrained within local power structures.

Impact of Climate Change on Conflicts:

As climate change progress worldwide, the situation in Darfur becomes evidently more critical. Erratic rainfall caused by climate change leads to droughts, reducing agricultural yields. Meanwhile, desertification increases as fertile land diminishes due to fluctuating weather patterns. As a result, tensions between Arab herders and African farmers escalate further as competition for resources intensifies.

Future Implications:

Unless proactive measures are taken immediately both globally and locally, climate change driven conflicts are only projected to worsen in Darfur. With expanding deserts encroaching on arable land, it threatens the livelihoods of both farmers and herders who rely on those resources for survival. Furthermore, mass displacement from uninhabitable areas may lead to expanded instability in the region - giving rise to extremist ideologies and even more organized violence.

Mitigating Climate-related Conflicts:

Global leaders must acknowledge the consequences of climate change on the international security landscape. Implementing climate adaptation strategies and providing assistance to vulnerable communities like Darfur would potentially reduce resource-driven conflicts. Additionally, investing in community-led initiatives, promoting resource sharing, and developing sustainable agricultural practices can help mitigate tensions in the region. Lastly, supporting regional peace-building efforts, addressing root causes of inequality, and promoting development can allow societies to better manage the impact of climate change.

Darfur is a prime example showcasing the intersection between climate change and resource-based conflicts. As global warming progresses at an alarming rate, it is crucial to address the environmental challenges that exacerbate existing socio-political conflicts in Darfur. Implementing practical solutions to these issues will not only contribute to global climate change mitigation efforts but also potentially create a more peaceful future in Darfur and beyond.

The Middle East, a region known for its volatile nature due to geopolitical tensions and resource-related conflicts, has been subjected to increased scrutiny. Over the years, nations around the globe have expressed concern about the long-standing issues plaguing this region. One emerging cause for anxiety is the evidence pointing towards the potential for wars induced by climate change. This essay will delve into the likelihood of wars breaking out because of climate change's effects in the Middle East, while considering political, historical and scientific aspects to provide a comprehensive analysis.

CLIMATE CHANGE AND Its Effects in the Middle East

To appreciate the connection between climate change and war, it is crucial to understand the impacts of environmental shifts on the Middle Eastern landscape. The primary climate-related issues that this region grapples with are water scarcity, desertification, rising temperatures, and irregular precipitation patterns. As an arid geographical location, water has always been a precious commodity which nations zealously guard. The exacerbation of water scarcity due to global warming has intensified tensions among states over resources sharing.

These effects have brought forth economic struggles and social instability within the populace as well. Farmers' livelihoods are threatened as fertile lands turn into deserts and crops fail; thus, their agricultural output diminishes. As economies falter under these pressures, governments are expected to invest in ensuring resource security within their territories.

———

THE HISTORICAL CONTEXT of Wars in the Middle East

To assess climate change's potential role as a catalyst for warfare in this region, it is essential to grasp past conflict contexts. Historically, wars and skirmishes have sprung forth from borders disputes between various factions over resources like oil and water sources (such as rivers basins). A prime example is the longstanding tension between Iran and Iraq over the Shatt al-Arab waterway.

Additionally, ethnic rivalries within countries like Syria or former political structures which have led to fragmentation (such as the aftermath of the Ottoman Empire) have contributed to the conflict spectrum. By connecting this historical understanding with the ongoing climate change-related strife, one can evaluate the risk of climate-induced warfare in the region.

Political Factors and the Role of Governmental Institutions

Governments and political entities play a vital role in either escalating or mitigating tensions fueled by climate change. As resources become scarcer, countries might opt to weaponize these challenges to further their interests within their borders and against neighboring states.

However, regional actors also possess the power to negotiate and engage diplomatically in strategies targeted at cooperative resource management. Agreements like the 1960 Indus Waters Treaty between India and Pakistan could serve as a framework for nations in water-stressed regions such as the Middle East. By reinforcing less combative methods and investing in collaborative endeavors, political bodies can successfully navigate through security challenges imposed by climate change.

Conclusion

The implications of climate change have brought various challenges for countries worldwide, but particularly for regions like the Middle East — where geostrategic competitions are commonplace. The potential for wars breaking out due to environmental shifts is a realistic and looming threat, especially if strategic politics continue unchecked. However, nations have the option to choose diplomacy over aggression in addressing water scarcity and resource-sharing concerns.

To evade this trajectory of war sparked by climate change effects, governments must invest in sustainable resource management practices that consider interconnected socio-economic relations within their nations, along with fostering regional cooperation for effective problem-solving. If collaborative efforts are prioritized over territorial instincts, then there is hope that an ecologically destabilized future does not inevitably lead to further conflict in an already strife-ridden region like the Middle East.

Chapter Sixteen

Strategies for Managing Climate Change and Migration

Climate change mitigation and adaptation measures.
Implementing green technology and practices. Infrastructure
projects to protect against climate impacts, seawalls, flood
defenses. International collaboration on climate change action
Developing policies and laws to accommodate economic
migrants

Climate change presents a wide array of challenges that both developed and developing nations must face in the coming decades. Among these challenges, migration driven by globalization, political instability, and environmental degradation has emerged as a critical issue. As a result, it has become increasingly important to consider the various strategies available for managing climate change and migration, which include mitigation and adaptation measures, the implementation of green technology and practices, infrastructure projects, international collaboration, as well as the development of laws and policies aimed at accommodating economic migrants. This essay will discuss each of these strategies in depth.

Mitigation and Adaptation Measures

Climate change mitigation refers to efforts aimed at reducing greenhouse gas emissions and enhancing sinks to remove carbon dioxide from the atmosphere. Examples of such efforts include investing in renewable energy sources, improving energy efficiency, afforestation, and land-use changes. By focusing on mitigating the factors contributing to climate change, these measures can better safeguard vulnerable populations that may be prone to displacement.

In contrast, climate change adaptation involves taking active steps to adjust human or natural systems in response to the effects of climate change. Some examples include building more resilient infrastructure, implementing early warning systems for natural disasters, improving water management systems, and promoting sustainable agriculture practices. Adaptation measures can be an essential component for managing migration driven by climate change, as they often address the immediate concerns and needs of affected populations.

Implementing Green Technology and Practices

Another vital strategy for managing climate change is the implementation of green technology and practices that reduce or eliminate greenhouse gas emissions and other pollutants during production processes. This encompasses a wide range of technologies including renewable energy production (wind, solar), energy-efficient transportation (electric vehicles), waste management systems (recycling), and agricultural practices (organic farming).

Infrastructure Projects

Another essential aspect of addressing climate change is investing in infrastructure projects that protect against its impacts. Such projects include building seawalls and flood barriers, restoring wetlands, improving drainage systems, and designing more resilient urban landscapes. These measures can help protect communities from the devastating effects of climate change and reduce the need for migration due to environmental reasons.

International Collaboration on Climate Change Action

Since climate change is a global problem, it is crucial for countries to work collectively on finding solutions. International collaboration on climate change action can involve sharing technology, finances, research findings, and expertise on mitigation and adaptation measures. Through increased cooperation among nations, it becomes possible to manage migration in a more orderly and sustainable manner, minimizing the negative impacts while maximizing potential benefits.

Developing Policies and Laws to Accommodate Economic Migrants

Lastly, it is essential for governments to implement policies and laws that address the issue of economic migrants affected by climate change. This might include offering temporary work visas or pathways to citizenship for those displaced by environmental conditions, adjusting labor market regulations to account for incoming migrants or advocating for international agreements that address the causes and consequences of environmental migration. By creating legal avenues for those affected by climate change to seek safety elsewhere, nations can reduce unauthorized migration and facilitate smoother integration processes.

MANAGING CLIMATE CHANGE and migration requires a multifaceted approach that addresses both the root causes of these challenges as well as their consequences. By implementing mitigation and adaptation measures, green technology, infrastructure projects, enhancing international cooperation, and developing responsive laws and policies regarding economic migrants, nations can confront the issue of climate-driven migration more effectively. Ultimately, such policies will not only benefit migrants but also contribute toward a more sustainable future for all.

Conclusion

A recap of the connection between climate change and economic migration. Emphasizing the need for global and regional cooperation on climate change initiatives. Call to action for readers to take part in addressing climate change issues

Climate change is undoubtedly one of the most pressing challenges faced by humanity today. The rise in global temperatures, increasing frequency of natural disasters, and rising sea levels continue to disrupt ecosystems and claim lives. One often overlooked consequence of climate change is economic migration. This essay will outline the connection between climate change and economic migration, emphasizing the need for global and regional cooperation on climate change initiatives. Lastly, a call to action will be presented for readers to take part in addressing this critical issue.

Connection between Climate Change and Economic Migration

As the impacts of climate change become more pronounced, many individuals are forced to leave their homes in search of a better livelihood. Economic migration is typically caused by a combination of factors such as declining agricultural productivity, water scarcity, loss of habitable land due to sea-level rise, and an increase in natural disasters like hurricanes and floods.

Agricultural productivity is particularly sensitive to changes in climate, as it directly impacts food production and availability. In regions where agriculture is the primary source of income, prolonged droughts or excessive rainfall can lead to crop failure, resulting in a loss of income and subsequent migration. For instance, farmer displacement in sub-Saharan Africa due to desertification has led to greater migration within the continent as well as Europe.

Similarly, low-lying coastal areas are facing significant challenges due to increasing sea-level rise and storm surges that threaten homes and critical infrastructure. Residents of small island nations such as Kiribati and Tuvalu face potential inundation of their home countries in the coming decades, leading to displacement on a mass scale.

The Need for Global and Regional Cooperation

The interlinked nature of both climate change and economic migration necessitates a comprehensive approach at the global level. Cooperation between different countries is essential in promoting a unified response that includes mitigation measures for vulnerable communities or social safety nets that can soften the blow from disaster events.

Firstly, collaboration on international climate agreements like the Paris Agreement is critical in working towards combatting global warming by reducing greenhouse gas emissions. Initiatives to transition to renewable energy sources and share technology or best practices can have a significant impact on mitigating the effects of climate change and reducing migration caused by it.

Secondly, regional cooperation plays a vital role in fostering resilience against climate vulnerabilities. Initiatives such as sharing early warning systems for natural disasters, coordinating humanitarian support during disaster events, or creating shared adaptation strategies can help countries work together to minimize the potential for climate-induced migration.

Call to Action

It is crucial for individuals to recognize that they are not powerless in addressing these global issues. Every person can take action in their daily life, from reducing carbon footprint through lifestyle changes like adopting a plant-based diet or using public transport to advocating for strong climate policies from government officials.

Moreover, raising awareness within communities about the consequences of human-induced climate change and encouraging participation in environmental movements or supporting clean energy initiatives will contribute towards the collective effort against global warming. Engaging in open conversations with friends and family about the link between climate change and economic migration helps spread information that can inspire positive change on a larger scale.

Climate change continues to exacerbate economic migration, presenting a growing challenge for affected individuals, communities, and countries around the world. Fostering global and regional cooperation on climate change initiatives will play a significant part in addressing this issue more effectively. Recognizing individual responsibility in mitigating these effects is necessary, driving collective action towards a sustainable future. By emphasizing these concerns today, we work towards preserving our planet for future generations while tackling challenging global crises such as economic migration.

In conclusion, the book has successfully managed to analyze the complex relationship between climate change and economic migration. By incorporating political, historical, and scientific perspectives, it becomes clear that the two global issues are intrinsically linked and require immediate attention. With migrants from the southern hemisphere increasingly seeking refuge in Europe, policymakers and governments must work to connect the dots between climate change impacts and the consequent economic migration.

Throughout history, human migration has often been driven by a need for resources or better living conditions. In recent years, we have witnessed a surge of people leaving their homes due to factors associated with climate change - such as droughts, floods, and other natural disasters. As global temperatures continue to rise and weather patterns become more extreme, affected populations in the southern hemisphere are bearing the brunt of these harsh environmental changes.

From a political standpoint, it is crucial for governments in both developed and developing nations to recognize that climate change is not only an environmental issue but also a humanitarian crisis. The influx of migrants from regions suffering from environmental degradation has become a significant challenge for European countries; as they grapple with increasing social tensions and limited resources for affected populations. Furthermore, leaving these regions depleted not only exacerbates existing inequalities but also puts them at greater risk of conflict over scarce resources.

By incorporating scientific research and evidence in our analysis, we can understand that growing evidence points towards climate change as a key driver of migration patterns around the world. Agrarian communities that rely on predictable weather patterns for crop growth face devastating losses when faced with intense droughts or sudden flooding events. This triggers economic hardship for families who have limited means to adapt and subsequently forces them into migration in search of better job prospects and living standards.

Therefore, it is imperative that we connect the dots between climate change impacts and the increase in economic migration from the southern hemisphere to Europe. This understanding will enable governments globally to develop policies addressing the root causes of migration while simultaneously tackling the environmental challenges related to climate change. Collaborative efforts, such as investing in sustainable development initiatives and promoting renewable energy sources in developing nations, will not only benefit the environment but also improve living conditions fostering socio-economic stability.

In summary, this book has elucidated the intricate relationship between climate change and economic migration, emphasizing the need for integrative and well-informed action on the part of global leaders. Acknowledging these issues as interconnected allows for the formulation of comprehensive policies that navigate societal and humanitarian challenges while fostering an environmentally sustainable future.

Appendix

Statistics and data related to climate change, migration trends, and economic costs

This section presents an appendix that compiles relevant statistics and data regarding climate change, migration trends, and economic costs. The data in this appendix stem from various political, historical, and scientific resources highlighting significant connections between these three phenomena.

Climate Change: Statistics & Data

1. Global Temperature Increase: The global temperature has increased by approximately 1.18°C (2.12°F) since the late 19th century (NASA GISS).

2. Glaciers Melting: Glaciers across the world are losing 267 billion metric tons of ice per year, contributing to sea-level rise (World Bank Group).

3. Sea-level Rise: Global mean sea level has risen by about 8 inches (20 cm) since 1900; projections for 2100 range from 29cm to 110cm (U.S. Global Change Research Program).

4. Extreme Weather Events: The number of climate-related natural disasters has tripled in the last 30 years (World Meteorological Organization).

Migration Trends: Statistics & Data

1. Environmental Migration: By 2050, approximately 200 million people could be displaced as a result of climate change impacts (International Organization for Migration).

2. Internal Displacement: In 2020, at least 30.7 million people were internally displaced due to natural disasters; more than three times the number displaced by conflict and violence (Internal Displacement Monitoring Centre).

3. Drought-induced Migration: Prolonged droughts have driven nearly two million internal displacements in East Africa between 2008 and 2018 (World Bank Group).

4. Island Nations at Risk: By mid-century, entire populations of low-lying island states such as Kiribati or Tuvalu could face forced relocation due to sea-level rise (United Nations High Commissioner for Refugees).

Economic Costs: Statistics & Data

1. Global Economic Impact: Climate change could reduce global economic output by 7.22% by 2100 if current emission trends persist (National Bureau of Economic Research).

2. Loss and Damage Costs: In the last two decades, climate-related disasters have cost USD 2.25 trillion, up from 895 billion in the previous two decades (Centre for Research on the Epidemiology of Disasters).

3. Infrastructure Damage: A 1-meter rise in sea levels could cause more than USD 1 trillion of damage to global infrastructure assets (OECD).

4. Agricultural Impact: Climate change impacts on agriculture could push food prices up by an additional 7-23% by 2050 (World Bank Group).

The appendix demonstrates that climate change, migration trends, and economic costs are intricately related phenomena with potentially devastating consequences. With increased political awareness and global collaboration, it is crucial to address these issues, mitigate negative effects, and develop proactive solutions to ensure a sustainable future for humanity.

References

C limate change is among the most pressing global issues of our time, with far-reaching implications for social, political, and economic systems. One ripple effect has been a significant increase in economic migration, or the movement of individuals and communities in search of new economic opportunities due to deteriorating living conditions resulting from climate change. This essay brings together a comprehensive list of resources on climate change and its impact on global economic migration patterns from political, historical, and scientific perspectives.

I. Political Resources:

1. Betts, A. (2013). Survival Migration: Failed Governance and the Crisis of Displacement. Ithaca: Cornell University Press.

2. Brown, O., & McGranahan, G. (2006). Environmentally Induced Population Movements and Environmental Impacts Resulting from Mass Migration. International Institute for Environment and Development.

3. Felli, R., & Castree, N. (2012). Neoliberalising Adaptation to Environmental Change: Foresight or Denialism? Environment and Planning A: Economy and Space, 44(1), 1-4.

4. Hartmann, B. (2010). Rethinking Climate Refugees and Climate Conflict: Rhetoric, Reality and the Politics of Policy

Discourse. Journal of International Development, 22(2), 233-246.

II. Historical Resources:

1. Adams H., Kay S., Nseluke Hambulo B.and Tschakert P.. (2018). Limits of Resilience? Exploring risk,response,and reprieve in Zambian farming systems.Global Environmental Change

2. De Haas H.. (2008) The myth of invasion: irregular migration from West Africa to Mediterranean EU _IMI Research Report_. International Migration Institute

3. Kniveton D., Smith C.D., Wood S., (2011). Agent-based model simulations of future changes in migration flows for Burkina Faso. _Global Environmental Change_, 21(1), S34–S40

4. McLeman, R., & Smit, B. (2006). Migration as an Adaptation to Climate Change. Climatic Change, 76(1-2), 31-53.

III. Scientific Resources

1. Adger, W. Neil. J Barnett & K Brown with N Marshall & K OH'Connell, (2013) Cultural dimensions of climate change impacts and adaptation Nature Climate Change 3:112-117

2. Cattaneo, C., & Peri, G. (2016). The Migration Response to Increasing Temperatures. Journal of Development Economics.

3. IPCC (2014). Climate Change 2014: Impacts, Adaptation, and Vulnerability. Part A: Global and Sectoral Aspects. Contribution of Working Group II to the Fifth Assessment Report of the Intergovernmental Panel on Climate Change.

4. Piguet E., Pécoud A., De Guchteneire P..(xxxx). Migration and climate change: an overview,_Revue européenne des`migrations internationales_ , IMI Working Papers Series 6–10

The resources listed above provide an excellent starting point for understanding the complex relationship between climate change and economic migration from various perspectives. As the ramifications of climate change become more apparent across the globe, it is crucial for researchers, policymakers, and other stakeholders to be well-informed about these issues in order to promote sustainable development and mitigate negative impacts on both human populations and the environment.

END

Economic Migration and Climate Change: The coming invasion (The Connection)

www.ingramcontent.com/pod-product-compliance
Lightning Source LLC
Chambersburg PA
CBHW031352160726
47993CB00002B/933